TROY THE TREE GUY!

Trees Rock! Growth, Selection, Planting, Care

Roger F. Hartwich Jr.   M.S.E., M.S.

TROY THE TREE GUY!: Trees Rock!

Growth, Selection, Planting, Care

First Print Edition

Disclaimer: This publication contains material primarily for educational and informational purposes. The author and publisher have made earnest effort to ensure that the information in this book was correct at publication time and do not assume and hereby disclaim any liability to any party for any loss, damage, disruption caused by errors and omissions.

Cover and Interior Design by Simon Goodway

ISBN: 978-1-7362828-3-0

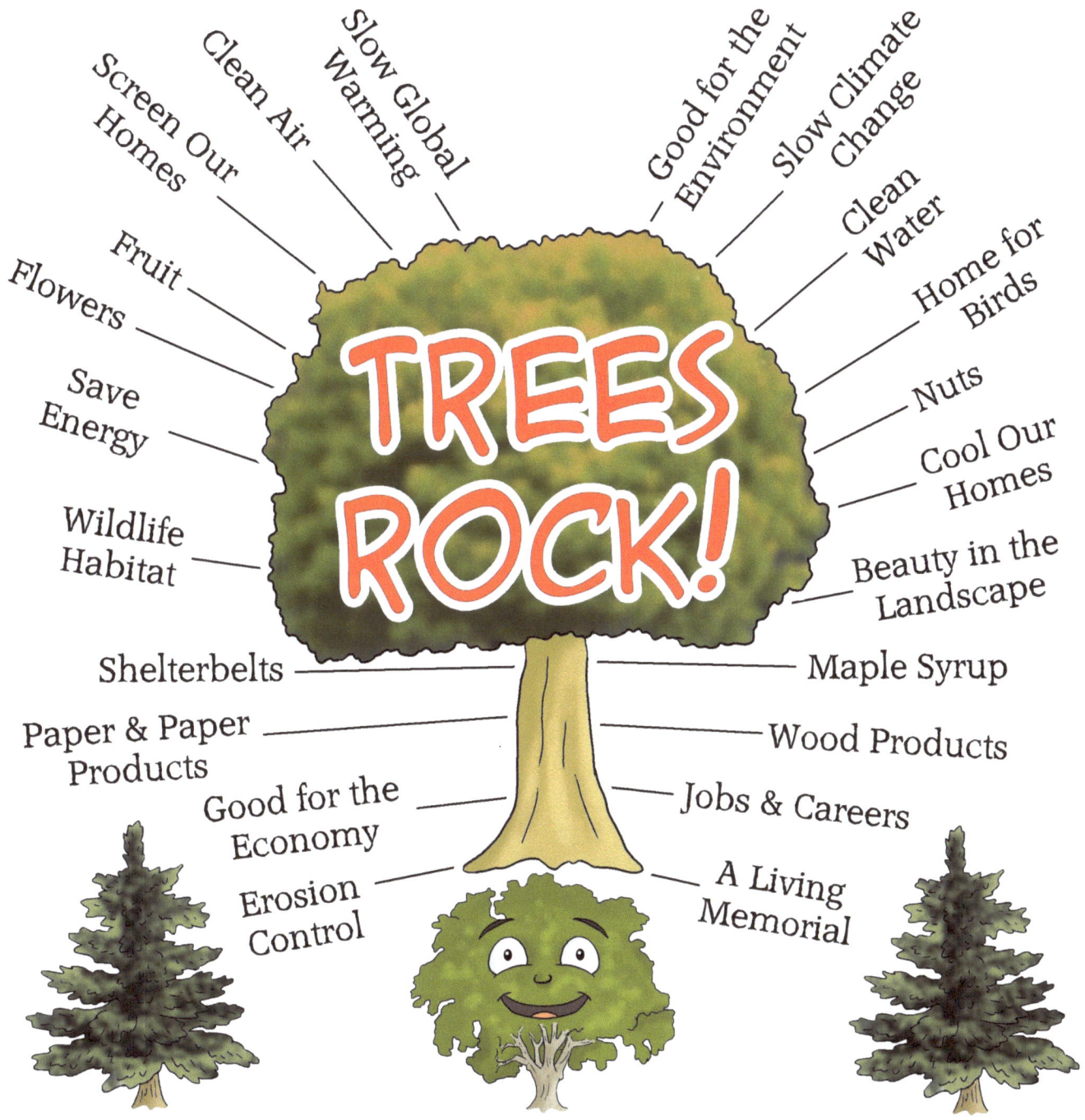

TREES ROCK!

- Screen Our Homes
- Clean Air
- Slow Global Warming
- Good for the Environment
- Slow Climate Change
- Clean Water
- Fruit
- Home for Birds
- Flowers
- Nuts
- Save Energy
- Cool Our Homes
- Wildlife Habitat
- Beauty in the Landscape
- Shelterbelts
- Maple Syrup
- Paper & Paper Products
- Wood Products
- Good for the Economy
- Jobs & Careers
- Erosion Control
- A Living Memorial

Table of Contents

Preface

To all people who love trees and realize the significant importance of trees to people, society, health, the economy, the environment, animals, birds, and other living things, this book celebrates trees and all that they stand for. This book is written and explained in simple terms, for middle to upper age youth and adults, based on fact, about tree anatomy, how trees live, grow, make food, and function, tree selection, and about their value and benefits to all of us. This book also explains planting, care, and maintenance of trees, including fertilization and pruning. Information regarding shrubs is also included.

Introduction

Have you ever wondered how trees grow, make food, function, and live? Do you know how to properly plant trees? Would you like to learn more about tree anatomy? About tree selection and about care and maintenance of trees and shrubs, including fertilization and pruning?

This book is for middle to upper age youth and adults about trees, based on fact, including tree anatomy, plant selection, planting procedures, and how trees live, grow, make food, and function. Information about trees is explained in simple terms, using a "bottoms-up" approach, from ground level to the top, from the seed, roots, trunk, to crown and points in between. This book also explains care and maintenance of trees, including fertilization and pruning. Care and maintenance of different kinds of shrubs is also included.

Hello. I'm Troy the Tree Guy. I live in a small town in the heart of the Midwest, near a lush green forest of beautiful trees. I grew up with a strong interest in plants, especially trees. During high school, I did volunteer work with our local garden club planting flowers, shrubs, and trees, then got a part time job at a shade tree nursery growing trees and shrubs. After high school,

I went to a university and double majored in horticulture and landscape architecture.

Currently, I own and operate a small nursery and garden center just outside of town. Our company grows plants and provides landscaping, landscape design, planting, tree and shrub pruning, and landscape maintenance services.

Since I have been in business for many years, I know a lot about trees, shrubs, flowering perennials, tall grasses, vines, and ground covers. In our nursery, we grow and pot most of our own plants to sell in the garden center, including trees and shrubs. We also sell balled and burlapped trees, and bare root trees in early spring. We specialize in growing shade and ornamental trees. We purchase some of our plants from other nurseries that we cannot grow or do not have space to grow them ourselves.

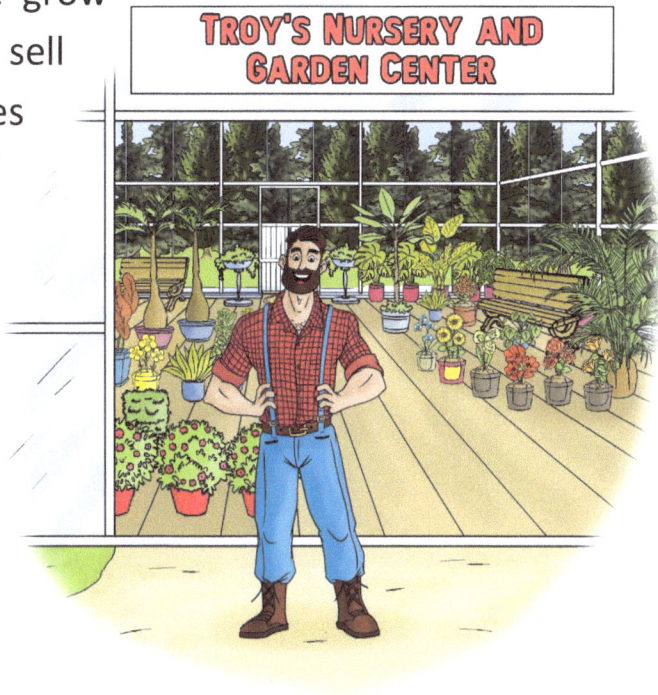

Today I'd like to talk to you about trees, their anatomy including the parts of trees, and how trees grow from seeds to **seedlings**, to become little trees, then to big trees. I will tell you how trees live, grow, make food, and function, and about how to select trees. I will also explain about care and maintenance of trees and shrubs, including principles of fertilization and pruning. Trees are very important to people, health, the environment, the economy, birds, animals, and all living things. Let's get started.

Kinds of Trees

Two main kinds of trees include **deciduous** and **evergreen** trees. Deciduous trees lose their leaves each year, usually in late fall. Evergreen cone bearing trees have needles that stay on the trees year-round. Examples of deciduous trees include maple, birch, elm, aspen, oak, linden, hackberry, northern catalpa, flowering crab, fruit and nut trees, and many others. Evergreen trees include pine, spruce, fir, and cedar trees. Many species and varieties of trees exist depending on the hardiness zone and area of the country in which you live.

Selecting Trees

Selecting appropriate trees for your yard or landscape should be done carefully as part of an overall unified landscape plan. The purpose and function of each tree should be kept in mind. Also, be sure that each tree selected is appropriate for your **plant hardiness zone** in which you live. The **plant hardiness zone** is the area or zone in which plants thrive best, based on the climate and temperature in the plant zone. Become familiar with the type of soil, the amount of sunlight needed, and moisture requirements in which the tree grows best. Also, look for trees that are disease and insect resistant.

Ask yourself key questions about trees that you may desire planting in your yard or landscape. Am I looking for a large shade tree to shade our home, helping to keep it cool and save energy during hot summer days? Am I looking for trees to filter out sunlight and shade our patio and picnic areas? Am I looking for trees to provide privacy screening from a busy road or from our neighbors? Am I looking for trees with beautiful spring flowers and brightly colored leaves to beautify the landscape around our yard? Am I looking for fruit or nut trees to provide nutritious food? Am I looking for trees that provide seeds, berries, or acorns for birds and animals? Am I looking for trees that provide habitat for wildlife?

Trees can be used for many purposes and functions. Plant trees that are hardy and grow well in your plant hardiness zone. Trees may provide fruit to eat, including apples, pears, peaches, plums, cherries, and others. Trees provide delicious nuts to eat, including walnuts, almonds, pecans, cashews, butternuts, hickory nuts, and others. Flowering crab trees provide aesthetic beauty with brightly colored spring flowers in shades of pink, red, and white. Trees may provide energy saving shade to help cool our homes from the hot summer sun, privacy screening, shelterbelts, protection from cold winter winds, erosion control, creation of outdoor living spaces, and many other functions.

Maple trees of different varieties provide brilliant orange or red leaves in fall, while other Maples provide dark purple leaves during the growing season. White Paper Birch trees provide beautiful white bark, while River Birch trees provide attractive, rust-colored, scaly bark. Gingko biloba trees, some of the oldest trees in existence, provide unique, fan-shaped leaves, are very resistant to disease and air pollution, and live a long life. Linden and Oak trees should also be considered. However, these are just a few examples of the many species and varieties of trees that can be used in a yard or landscape.

The following is a partial list of deciduous trees of various sizes that you may be interested in learning more about. Your local nursery and garden center may be able to give you more information about these and other trees, depending on the plant Hardiness Zone and part of the country that you live in with various climates and temperatures.

Large shade trees: Norway Maple, Schwedler Maple, Royal Red Maple, Crimson King Maple, Red Maple, Sugar Maple, other Maple varieties, Alder, Buckeye, Oak, American Linden (Basswood), River Birch, white Paper Birch, Beech, Thornless Honeylocust, Aspen, Hackberry, Northern Catalpa, Poplar, Willow, and others, depending on the plant Hardiness Zone that you live in.

Small to medium size trees: Amur Maple, Shadblow Serviceberry, Eastern Redbud, Hawthorn, Ginkgo biloba, Little Leaf Linden, and others, depending on the plant Hardiness Zone that you live in.

Ornamental trees: Flowering crab, Smoke tree, Amur Cherry, Peegee Tree Hydrangea, Canada Red Cherry, European Mountain Ash (with bright orange berry clusters), Japanese Tree Lilac, Pagoda Dogwood, Serviceberry, Amur Maple, Tatarian Maple, Nannyberry Viburnum (with spring single white flowers and late summer blue-black fruits to attract birds), Willow, Magnolia, and others, depending on the plant Hardiness Zone that you live in.

Fruit and Nut Tree: Fruit and nut trees may include apple, cherry, peach, pear, plum trees, and others, depending on the plant Hardiness Zone that you live in. Check the tree's hardiness zone in which it grows best.

Evergreen Trees

Evergreen trees should be highly considered for use in a yard or landscape. They may include pine, spruce, fir, and cedar trees. Pyramidal evergreens including pyramidal junipers and pyramidal arborvitae are often used as screens and hedges.

Evergreen trees can be used in combination with deciduous trees to provide beauty and function. They may be used to screen out busy streets and roads, neighboring homes, buildings, and unsightly views. Some evergreen trees come in a variety of colors, such as Colorado Spruce trees with dark green or blue-green needles. Balsam and Fraser fir trees make excellent holiday Christmas trees. Evergreen trees may also function as shelterbelts, windbreaks, and perimeter plantings. Evergreen shrubs are also used in combination with deciduous shrubs for good

contrast of color, texture, year-round persistent green, and fewer maintenance requirements.

The following is a partial list of **evergreen trees** to consider:

Red (Norway) Pine, White Pine, Scotch Pine, Jack Pine, Mugho Pine (shrub or small tree), Norway Spruce, White Spruce, Black Hills Spruce, Colorado Spruce (blue, green), Douglas Fir, White (Concolor) Fir, Balsam Fir, Fraser Fir, Pyramidal Juniper varieties, Canadian Hemlock, American Arborvitae, other Arborvitae varieties, and others, depending on the plant Hardiness Zone and part of the country that you live in.

Seeds

Growth of a seed starts with a tiny **embryo** inside the seed. Inside the seed, this tiny **embryo** contains cells that form tiny leaves, a stem, and a point that will become the root. First, the embryo cells expand, and the **embryo** splits the shell or outer coat of the seed, called **germination**. Then the root probes or searches its way to the ground and goes into the soil.

The **root** starts absorbing or taking in water and mineral nutrients. After that, leaves will emerge from the shell, capture energy from the sun, and create a supply of **chlorophyll** (the green pigment found in **chloroplasts**

in the leaves). This will allow the tree to manufacture its own food from water in the soil and carbon dioxide in the air, a process called **photosynthesis**.

Some nursery and garden centers purchase and start with **seedlings** to plant in pots. **Seedlings** are young plants that are grown from seeds.

Roots

Roots are at the bottom of trees and, as explained above, are very important to the growth of a tree. Roots grow where moisture and oxygen are available. Roots store water and mineral nutrients. They are found in the top three feet of soil, most in the top 18 inches, and extend out beyond a tree's **dripline.** A dripline is the outer part of a tree's crown. A **crown** is the top part of a tree. A crown, containing leaves and branches, gives a tree form and shape at the top. Trees of different kinds and varieties may have different shapes and sizes.

Large roots, called **taproots,** anchor trees to the ground and help them stay upright and tall during strong winds. During heavy rains or floods, roots help to hold the soil in place to prevent erosion, the washing away of the soil. Large roots also store water and mineral nutrients and help carry them up into the tree. **Smaller roots**

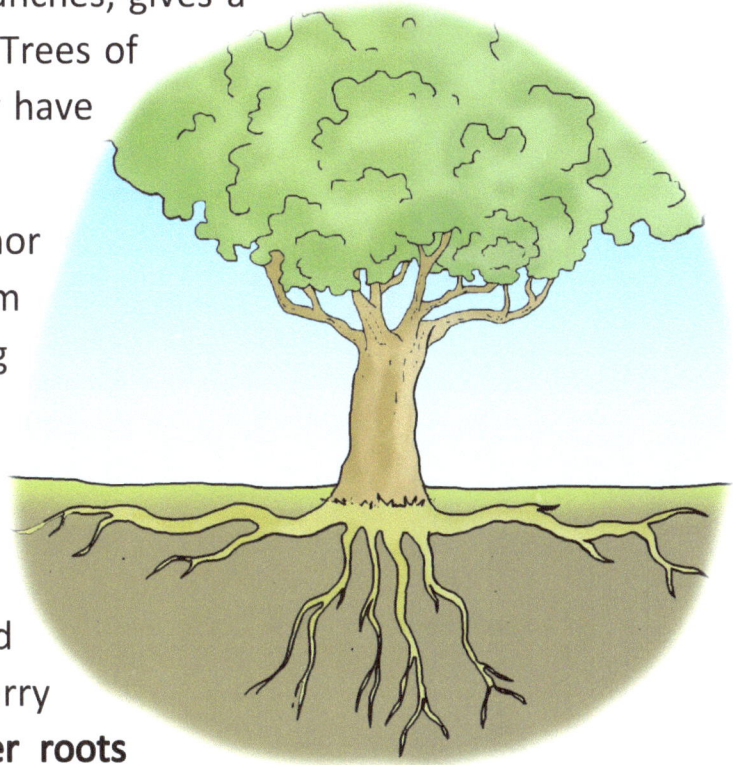

called **fibrous roots** take in water and minerals from the soil. Most of these smaller roots are found closer to the earth's surface in the upper 12 inches of soil.

Trunk

Next comes the tree **trunk**. This is the largest part of a tree. A trunk is made up of several different layers, all very important to a tree's life, growth, and health.

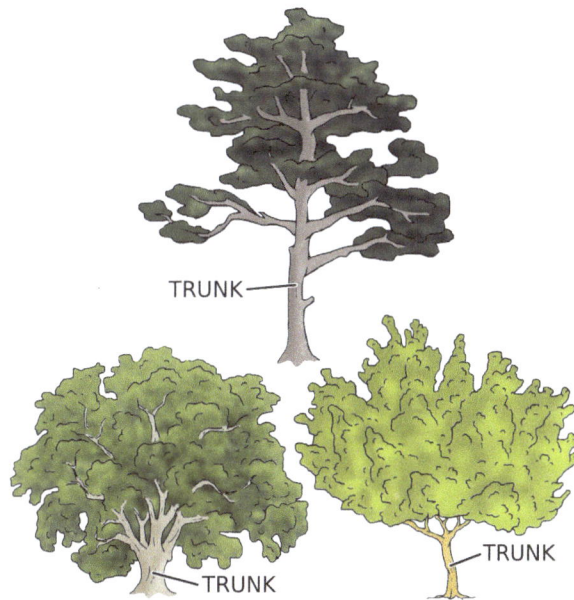

Outer Bark

Bark is the outermost part or layer of the tree trunk, stems, and branches. This layer protects or "insulates" a tree from the outside world, including bad weather and storms. When the air is very hot and dry from the hot sun, bark helps to control the temperature inside the trunk, stems, and branches. Bark prevents the tree from losing moisture and water from the trunk.

Bark also helps to protect a tree from injury. Bark on some trees is very attractive and adds beauty to a landscape, such as white bark on white Paper Birch trees, lustrous, reddish-brown peeling bark on River Birch trees, reddish-brown scaly bark on Eastern Redbud trees, or smooth, blue-gray bark on Blue Beech trees.

Cambium Layer

The next layer of a tree is the **cambium cell layer**. This layer is where growing happens, from growth of a small tree to a large tree. Every year, this layer produces new bark and new wood. The **cambium cell layer** produces two kinds of tissue that will transport water and minerals to the tree. These tissues include the **xylem or sapwood** on the inside of the cambium layer, and the **phloem** on the outside of this layer nearest the outer bark.

Phloem

The **phloem** is called the inner bark of a tree. The phloem is a tube or vessel like a pipeline. Food, including carbohydrates (sugars), is passed **downward** from the leaves through the phloem, and in other directions to other parts of a tree. The phloem lives a short period of time. It then dies, turns to cork, and becomes part of the outer bark of a tree, which helps to protect the tree.

Xylem or Sapwood

The **xylem, or sapwood**, is on the inside of the cambium layer, more toward the center of the tree. The xylem or sapwood is a tube or vessel like a pipeline. Instead of moving food downward as in the phloem, water moves **upward** to the leaves through the xylem.

The **xylem or sapwood** is new wood of a tree. Newer rings of sapwood are laid down on the outside of the wood, and the inner cells turn to **heartwood**, the fifth and center-most layer of a tree.

The **xylem or sapwood** has several other important jobs to do for a tree. Besides bringing up water, the xylem or sapwood also brings up minerals and nutrients from the roots to leaves and other growing points of the tree. The sapwood helps to defend the tree against the spread of disease and decay. Sapwood is very strong and helps to support the weight of a tree.

Heartwood

The **heartwood** is the fifth layer of a tree, the farthest layer inside a tree. The heartwood is the central or innermost part of the xylem. The heartwood is very strong. It contains no living cells and may be darker in color than the sapwood. It will not decay or lose strength when it is protected by the outer layer of a tree. Some trees do not have heartwood.

Stems: Branches and Twigs

Branches and **twigs** grow out of a tree trunk. They are attachment points and support leaves, flowers, and fruit that grow on the tree. Stems also help to form the structure of a tree.

Branches support or carry twigs. Twigs are small stems that support the leaves, flowers, and fruit. **Buds** may occur along a twig, at the base of each leaf, under the bark, or at the tip of each twig. Two of the main kinds of buds are the **terminal buds** at the end of a shoot, and the **lateral buds** along the stem. These buds may develop into a flower or **shoot**.

The **terminal buds** are usually the most active on branches or twigs. The **lateral buds** may be inactive or dormant due to the dominance of the terminal buds which prevents growth and development of the lateral buds. When **pruning** removal is done to the terminal buds, the dormant buds may become active and new shoots may develop.

Branches are strongly attached to the wood and bark below the main branch. However, branches are weakly attached to the wood and bark above the main branch. Branches and twigs help to give a tree, as well as shrubs, form and shape.

Leaves and Photosynthesis

Leaves are vitally important to the life of a tree. Leaves make food for a tree, but how? Let's take a closer look.

Leaves have cells with **chloroplasts**. Chloroplasts contain a **green pigment** called **chlorophyll**. A pigment gives something its color.

Next, the green pigment **chlorophyll** in a leaf absorbs, or takes in, sunlight. Energy from the sunlight is collected in the chloroplasts in the cells. In these cells containing chloroplasts, a reaction takes place called **photosynthesis**. The term photosynthesis can be broken down into two words **"photo"** meaning light, and **"synthesis"** meaning to put together.

The energy from the sunlight is converted to chemical energy in the form of **carbohydrates**. Water and carbon dioxide are necessary for photosynthesis to happen. A tree absorbs carbon dioxide from the atmosphere through the **stomata**, small openings on the undersides of leaves and other green plant parts. Light energy is changed to chemical energy (carbohydrates combining carbon, hydrogen, and oxygen to form sugars and starches) used for growth of a tree.

To review, **photosynthesis** takes place in cells that contain **chloroplasts**. These chloroplasts contain the green pigment called **chlorophyll** that absorbs or takes in sunlight and gives off oxygen into the air. During photosynthesis, leaves make food for the tree, and release oxygen into the air through the stomata. **Photosynthesis** may read like this: Water + carbon dioxide + energy (sunlight) in the presence of chlorophyll yields or gives off oxygen and sugar.

Many factors affect photosynthesis. These include the intensity or brightness of the light, length of time of the light, temperature, and water available to the plant.

Tree Growth

Tree growth includes the roots getting established the first few years. Trees get taller, wider, and thicker during their entire lifespan. This is in contrast with humans who stop growing when grown up. Trees do not grow from the bottom up as many plants do. Trees grow from the top up and on the outside of a tree. Trees continue to grow their outer shell which contains the living parts of the wood. This includes the **xylem**, which moves water up from the roots, and the **phloem**, which moves food down from the leaves. Trees continue to grow these outer, living shells of wood to stay alive.

You may wonder about tree rings and what they tell us. During spring and summer, a tree adds new layers of wood to its trunk. Spring wood grows faster than summer wood. Spring wood is light color with larger cells, and summer

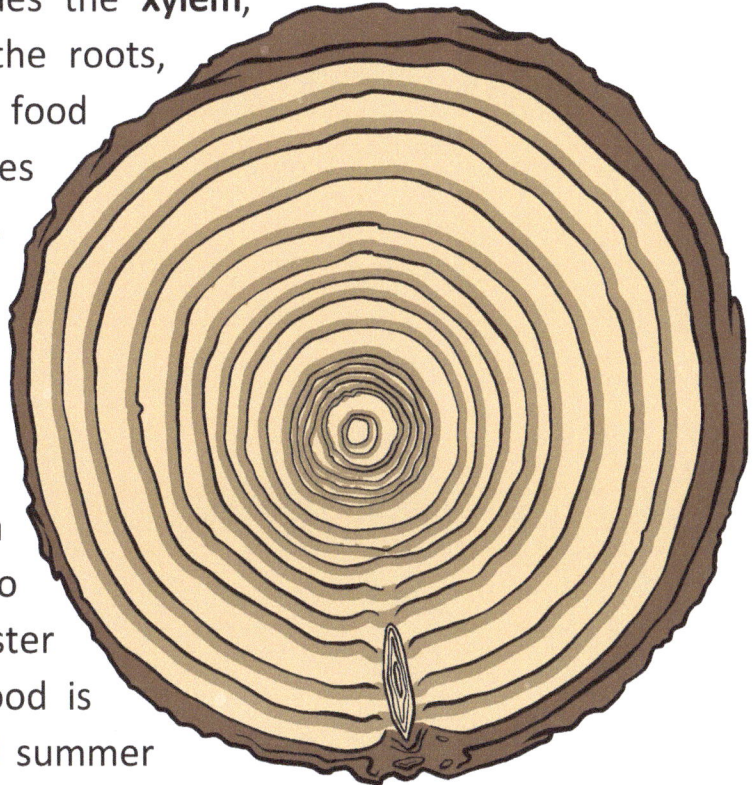

growth is slower and darker with smaller cells. You can tell a tree's age by counting the dark rings. Water, sunshine, and other environmental conditions can affect the growth of a tree, such as forest fires and drought. Insect infestations can also adversely affect the growth of trees.

Crown

The **crown** is the uppermost part of the tree. It is made up of leaves, stems including the branches and twigs, and berries and flowers on many trees. On fruit and nut trees, the crown contains the fruits and nuts.

The **crown** is made up of various sizes and shapes, depending on the kind of tree. Large shade trees usually have large crowns which help to shade or screen homes and buildings. Smaller ornamental trees, such as flowering crab trees, have smaller crowns, some rounded, some more pyramidal in shape. Fruit and nut trees have crowns of various sizes and shapes.

Tree Planting

When planting and growing trees, it is very important to match the kind of tree to the planting location, including its needs, requirements, and preferences for best growth and development. These include type of soil, for example clay loam or sandy soil, more acidic or basic soil, wet or dry soil or in-between, good drainage, sun, semi-shade, or shade. When purchasing a tree, look for one that looks healthy, with good leaf color, good form and shape of the crown, good symmetry, growing upright and straight.

Four types of nursery stock that can be purchased include **bare root**,

containerized, container grown, and balled and burlapped (B&B) trees. Most of the information presented here will include containerized, container grown, and balled and burlapped trees. These are the primary ways that trees are sold in nurseries and garden centers.

Bare-root trees are planted with the main structural roots near the surface of the soil. Roots are spread and distributed laterally near the soil surface to prevent kinking or circling of the roots. During planting, the planting hole is gradually backfilled, making sure the plant stays upright and straight. After planting bare-root trees, water and saturate the planting hole area. Staking may be required.

Some trees are **container-grown** in the nursery and sold that way in a garden center. With container-grown plants, roots are more established in the soil in the pot with a layer of material below the soil surface called **substrate.** The **substrate** may contain mineral nutrients and holds moisture to support the growth of the plant. Other plants are grown in the nursery field, potted in containers, held for several weeks or months, then sold in containers without established root systems.

Planting potted plants requires careful procedures and attention. Remove most containers before planting unless the container is biodegradable. Even biodegradable containers may be removed if the root system and surrounding soil stay together. Check the root system for any circling roots which should be separated and spread outward. Some far-extending roots may need to be cut when planting. Separating and spreading out the roots will prevent root girdling. Water thoroughly after planting and make sure soil is compacted around the roots to prevent air pockets.

Balled and burlapped trees in the nursery are dug with some of the root ball intact and wrapped with burlap to keep the soil in place. It is essential to remove the top and upper side portions of the burlap even with biodegradable burlap. This prevents wicking away of moisture from the plant. All synthetic burlap should be removed. Remove all twine, natural or synthetic, from around the trunk. Large balled and burlapped trees may come in wire baskets to hold the ball together. For traditional wire baskets, remove as much of the top part of the wire as you can to allow roots to grow and spread freely near the surface. This should be done after the tree is placed in the planting hole and stabilized with backfill.

Planting Procedure Summary

To review, most vigorous root growth occurs near the surface of the soil. Root growth from the bottom parts of the roots may be slower due to inadequate soil aeration and poor drainage. It is best to have the planting hole larger than the root ball, perhaps twice or more width of the root ball at the soil surface and less at the base of the root ball. This allows the root system to grow more easily and quickly into the surrounding landscape soil. This pertains to planting a potted plant as well.

When planting, be careful not to plant too deeply, never deeper than the **trunk flare** to the bottom of the root ball. In other words, the **trunk flare** is just above the soil surface and not buried. Planting a tree too deeply may suffocate or stress roots, or even drown the roots if planted in soil with poor drainage, such as in heavy clay soil. For younger plants, locate where the roots spread from the stem on the plant, called the **root flare**, and plant at or just below the surrounding level.

When planting a potted plant, remove the plant from the pot, making sure to keep the soil around the root system intact with the soil firmly around the roots. In some cases, not always recommended and not always the best, with a wooden biodegradable pot, a plant can be planted with the pot intact to keep the soil together around the roots. This would be done if the pot is made of wood which will biodegrade in time and allow the roots to expand and grow into the surrounding soil.

Use the same natural surrounding soil in the landscape to backfill around the tree to maintain the same soil texture. You may add water as you backfill the soil around the roots to eliminate air pockets. Keep soil around the bottom of the root ball firm as you add soil. Tamp down the soil as the hole is being filled to keep the tree vertical. Water the tree slowly but thoroughly after backfilling. Regular watering should be done thereafter, checking to make sure the tree is growing straight and tall. Mulch with hardwood or similar bark around the tree trunk to keep the soil cool and moist, preventing the soil from drying out. In some cases, the tree may require staking and a tree wrap may be recommended.

Tree wraps are used during winter months to protect young trees with thin bark from **sunscald**. Some trees with thin bark include maples, linden, gingko, crabapples, poplar, aspen, sycamore, and Japanese maple. **Sunscald** usually occurs during winter on the south or southwest side of a tree. The sun hits the tree, warms the cells in the cambium, and causes the cells to move water and nutrients. At night, the temperature drops, the cells may freeze, burst, and cause the bark to split. Tree wrap should be removed in early spring. It is recommended that trees with thin bark should only be wrapped for the first one to three years following planting, but opinions do vary.

Plant Care and Maintenance: Fertilizing and Pruning

Proper and timely tree care and maintenance are very important for the growth, development, and health of a tree, as well as for evergreens, shrubs, and other flowering plants. This includes appropriate fertilizing and pruning of trees and shrubs.

Fertilizing

Most tree roots are found within the top three feet of the soil surface, the most vigorous within the top 18 inches. These roots extend out to the crown of the tree, or a short distance beyond the crown. Young trees planted may not need fertilizing the first year or two because the roots are getting established. It is best to fertilize trees in late August through September. Early spring before new growth starts is also a good time. Avoid fertilizing on hot, dry summer days during times of drought stress on trees. Adequate water is needed for plants to absorb nutrients.

Tree fertilizing may be beneficial and necessary in some landscape situations for satisfactory development and growth of trees. This includes providing **nutrients** or **essential elements** for landscape trees that are growing in soils that do not contain the proper nutrients. Essential **nutrients** are taken up by the tree roots

Three primary nutrients, called **macronutrients**, include **Nitrogen**, **Phosphorous**, and **Potassium**. Each nutrient performs one or more primary functions. When you purchase fertilizer, these nutrients will be listed on the container with the percentage of each nutrient listed.

Nitrogen is mainly responsible for the growth of leaves in the plant and is often required in the largest quantity. Nitrogen is very important to photosynthesis, green growth, and flowering.

Phosphorous is mainly responsible for root and stem growth. **Potassium** helps the overall functions of the plant.

Different types of fertilizers, organic and inorganic, are absorbed by the plant at different rates. **Organic** materials come from plant or animal sources, such as composted plant materials, and release nitrogen into the soil. Nitrogen is then taken up by plant roots.

Other fertilizers, called **inorganic** fertilizers, include commonly used turf fertilizers. These fertilizers may have an outer coating that breaks down slowly to release the fertilizer. Slow-release fertilizers release nutrients, especially nitrogen, into the tree over a longer period. Slow-release fertilizers save time, can be applied quickly, and can prevent burning of turf or roots of plants. In

many cases, lawn fertilizer applied to turf grass areas around trees is sufficient.

Proper water or moisture levels around the tree roots is essential for satisfactory uptake of the nutrients into the tree. Fertilizers can be applied in several different forms including granules which are absorbed into the soil, tree spikes put into the soil in concentric circles around the tree (not recommended by some individuals), liquid applied to the surface of the soil, or liquid applied under the soil surface using drill holes in the soil. Liquid fertilizer can also be injected into the soil under high pressure.

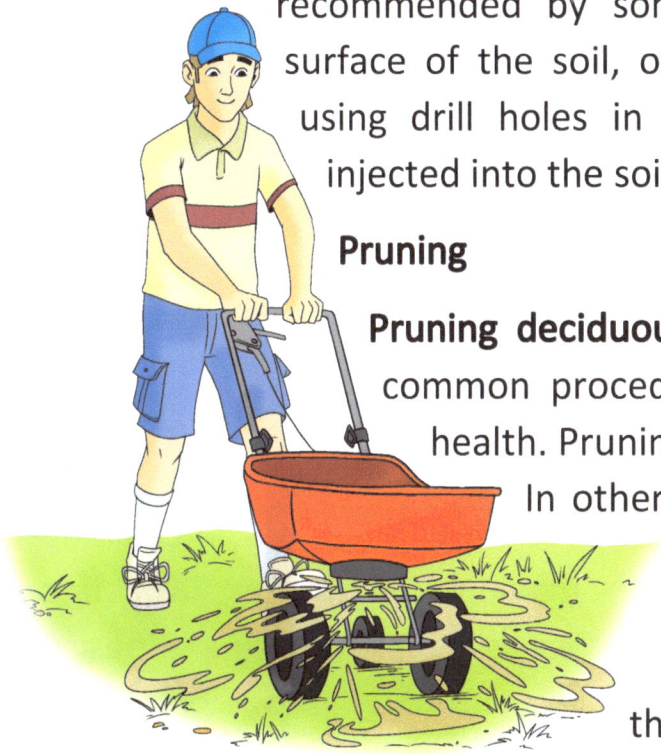

Pruning

Pruning deciduous trees, shrubs, and other plants is a common procedure to maintain a plant's vitality and health. Pruning should be done with purpose in mind. In other words, it is important to have a good reason or reasons for the pruning of a limb, branch, or branches. These may include safety, improvement of the structure and shape of a tree, and for the tree's overall health and aesthetic appearance. Pruning trees at the time of planting should be limited to removal of broken, dead, and damaged limbs. Minimal pruning may also be recommended to remove a double leader or to reshape the plant.

Pruning can be done to train or shape a plant, control plant size, and for better appearance, health, and rejuvenation of a plant. Plants can also be pruned for special uses such as shade, windbreaks, or sound barriers. Fruit trees, such as apple trees, should be pruned appropriately for better and more healthy fruit production.

Low tree branches over sidewalks, driveways, and roads should be pruned for pedestrian and vehicle clearance safety. Other reasons for pruning include removal of dead, weak, broken, or undesired tree branches over homes and other buildings, reducing undesired shade over buildings, and improving a view. Pruning in overplanted or overcrowded areas to allow for improved air circulation and light penetration is another important consideration.

It is important to know the growth habits of your plants, the structure and shape of each plant, their mature size, and what you want them to do for you. Tree pruning principles include removal of dead and dying limbs, stems, branches, and twigs which can be done any time. Removal of overcrowded and crossing branches to prevent rubbing, to create more space to allow light penetration and air circulation in and around the tree, is also recommended, especially on fruit trees for healthy and improved fruit growth.

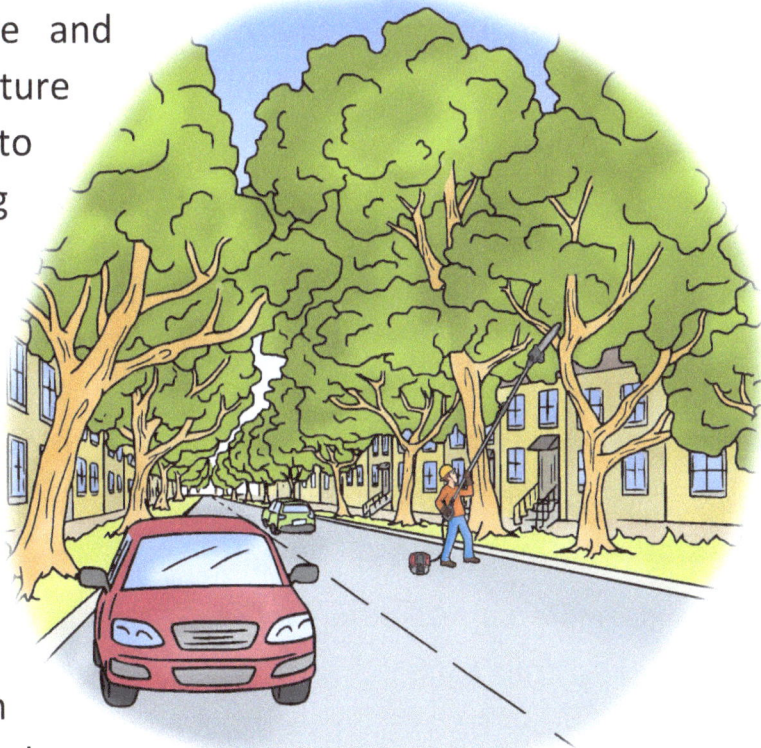

Timing of Pruning

The best time to prune trees, as well as shrubs, varies depending on the kind of tree, the condition of the plant, its growing season length, climate in different areas of the country, when it blooms, and whether its' buds form on new growth or the previous year's growth. The best time to prune trees is normally during the winter months, approximately November through March. This is when many trees are **dormant** or inactive and less susceptible to disease if they are pruned during this time. Trees can then heal more quickly in the spring. It is best to prune many shade trees in late summer or early fall. Some trees, such as the ginkgo biloba, are very slow growing and require little pruning.

It is best to prune oak and maple trees during winter. Some recommendations advise pruning of maple, birch, and weeping willow trees in late summer or early fall, not in late winter or spring since these trees may bleed severely if pruned in late winter. However, a small amount of sap dripping or bleeding from pruning will not harm a tree. The cut will heal in time. Excessive sap dripping over a long period of time may lead to the tree being susceptible to pests and disease, poor tree health, and may lead to the death of a tree. The sap carries water and nutrients through the tree to keep the tree healthy. Dead, weak, broken, and dying branches can be removed at any time during the year. Most pruning of flowering trees, including flowering crabs, should be done after the blooms fade, but this can vary. Trees that produce berries or fruit should preferably be pruned in late winter.

Minimal pruning at the time of planting is recommended. However, young trees should be pruned to establish a good structure to minimize pruning when the tree is older. This includes weak branch attachments and double or codominant stems or trunks. The central stem or trunk is the leader. If two

stems or trunks exist, the most vertical and strongest stem must be left in place while the other less dominant stems or trunks are removed or cut back severely. In future years, competing stems or trunks should be cut back allowing the remaining central leader to grow straight and vertical. When needed, pruning one to three years or more after planting to produce a strong and balanced branch structure is recommended.

Pruning Evergreens

Evergreens include narrow-leafed and broad-leafed evergreens that require different kinds of pruning. **Narrow-leafed evergreens** include fir, juniper, pine, spruce, yew, cedar, arborvitae, and hemlock trees. **Broad-leafed evergreens** include azalea, rhododendron, boxwood, holly, camellia, laurel, and others. Evergreens make their new growth from buds that were formed the preceding year. Broad-leafed evergreen pruning should normally be limited to removing fading flowers shortly after they have bloomed.

Shearing, a type of pruning, is cutting off new growth of a tree or plant during summer to confine, hinder, or slow the growth of a plant. Shearing involves cutting off the ends of all branches on the outside of a deciduous or evergreen shrub or evergreen tree. This may promote a thicker, denser plant. In contrast, ordinary pruning is removal of woody limbs, including removal of diseased, dead, broken or dying branches.

Narrow-leafed evergreens, including pine, spruce, fir, and others listed above, are most often sheared than pruned. Shearing evergreens is best done in late summer or fall. Evergreen pruning of limbs, only if needed, should preferably be done in late summer, fall, or winter when the tree is dormant to avoid bleeding.

When pines grow, they send out many little shoots called **candles.** When the candles are about two to four inches long, they can be pinched or pruned off at the very tips to slow the growth of the tree.

Pruning Shrubs

With shrubs, proper timing of pruning depends on the various kinds, species, and varieties of plants and is critical for bud and flower development. Many shrubs that bloom during the summer on the current year's wood should preferably be pruned at the end of the summer after the plant has flowered.

Some plants bloom on year-old wood, including lilacs, forsythia, weigela, rhododendron, azalea, spring-blooming spirea varieties such as bridal wreath, and other plants. The best time to prune these shrubs that bloom on year-old wood with year old buds is just after the blossoms have faded. The shrub will then grow new branches and form buds that will bloom the following year.

An exception to the above pruning time for those plants that bloom on year-old wood includes very large, overgrown plants that need more immediate pruning. This immediate pruning may be appropriate if the purpose is to control the size of an overgrown plant, to improve its overall health, rejuvenate the plant, and not be concerned about losing some potential flowers for the following year. Removing all dead, weak, dying, and broken branches should be done as well.

Pruning to allow for air circulation and light penetration into and around a plant is important for proper flower and fruit growth. Thinning, the removal of a certain number of the main stems of a shrub, may be necessary when a plant has been neglected or allowed to grow too large without pruning for several years. It is advisable to only remove one third of the main branches of a plant each year. If a plant is extremely overgrown and needs extensive thinning and pruning, it may be wise to thin or prune back the plant more severely for optimum future plant growth and development. Some shrubs do best if severely pruned every one or two years.

Pruning Summary

To review, timing should be kept in mind when pruning. Removal of dead, broken, weak and dying branches can be done any time. Most trees should be pruned during the winter months when they are dormant and less susceptible to diseases. They then can heal more quickly in the spring. Pruning of oak and maple trees during the winter months is best. Apple and other fruit tree pruning is also best during the winter months, or before buds and flowers develop in early spring.

When pruning and removing stems, branches, twigs, and major limbs, it is essential to have purpose and objectives in mind. Pruning in a systematic, well-thought-out manner is critical for the plant's overall growth, development, and health.

Trees: People, Environment, Food, Birds and Animals, Economy

As stated before, I live in a small Midwest town. Nearby is a large, deep forest with lots of beautiful trees and many kinds of plants and animals. Trees living in this forest include oak, maple, elm, birch, aspen, ash, walnut, and hickory trees. Some of the evergreen trees include Norway or Red Pine, White Pine, Jack Pine, Norway Spruce, White Spruce, and Black Hills Spruce.

Trees in the forest and in our yards, parks, towns, and cities are good for the environment, for people, health, birds, and animals. Trees absorb or take in carbon dioxide and give off oxygen for living things to breathe. Trees help to provide clean air and filter air pollutants. Trees help to keep water clean in the lakes, rivers, and streams around us. Trees help to slow global warming and climate change.

For many birds and animals, trees provide food, shelter, and homes. Acorns from oak trees provide food for squirrels and other animals, and berries provide food for birds and animals. Evergreen trees provide habitat, food, and shelter for many animals. Trees provide food for people including fruits and nuts, and lumber for our homes, furniture, and other products made of wood. Trees provide paper and paper products for our offices, businesses, schools, and homes.

Trees are good for the economy and provide many careers and jobs in fields related to trees. These occupations include foresters, loggers, lumber yard workers, tree nursery growers, garden center owners and workers, carpenters, paper mill workers, landscapers, landscape designers, landscape architects, furniture store owners, cabinet makers, salespeople, and many others. Trees are of great benefit to people, health, the environment, animals, birds, and all living things.

Trees provide beauty in the landscape,
whether it be in a family's front or back yard, a city, state, or national park, or along roads, streets, and highways. Trees provide privacy screening from neighbors or busy streets, property boundary lines, shelterbelts, windbreaks, shade for our homes to save energy from the hot summer sun, and noise

absorption from noisy traffic on busy roads and highways. Ornamental trees provide bright beautiful flowers during spring, and many shade trees provide lustrous, colorful leaves during the fall season. Fruit and nut trees provide healthy, delicious foods.

Summary/Conclusion

It is hoped that you, the reader, have learned important things about trees, how they live, grow, make food, and function, and how to select and care for trees. It is also hoped that you realize the great value and wonderful benefits of trees.

All trees use the process called **photosynthesis** to make food. Trees take in carbon dioxide gas and give off oxygen for people and all living things to survive.

What can you do? Study and learn all you can about trees. Learn how to plant trees, then plant trees with a friend, family member, or others interested in trees. You will then be doing your part in improving our health and beautifying our

45

environment with trees for the present generation and generations to come. Visit your local nursery and garden center, arboretum, and parks to learn more about trees, and about other plants including shrubs, annual and perennial flowers, tall grasses, ground covers, native plants, and vines.

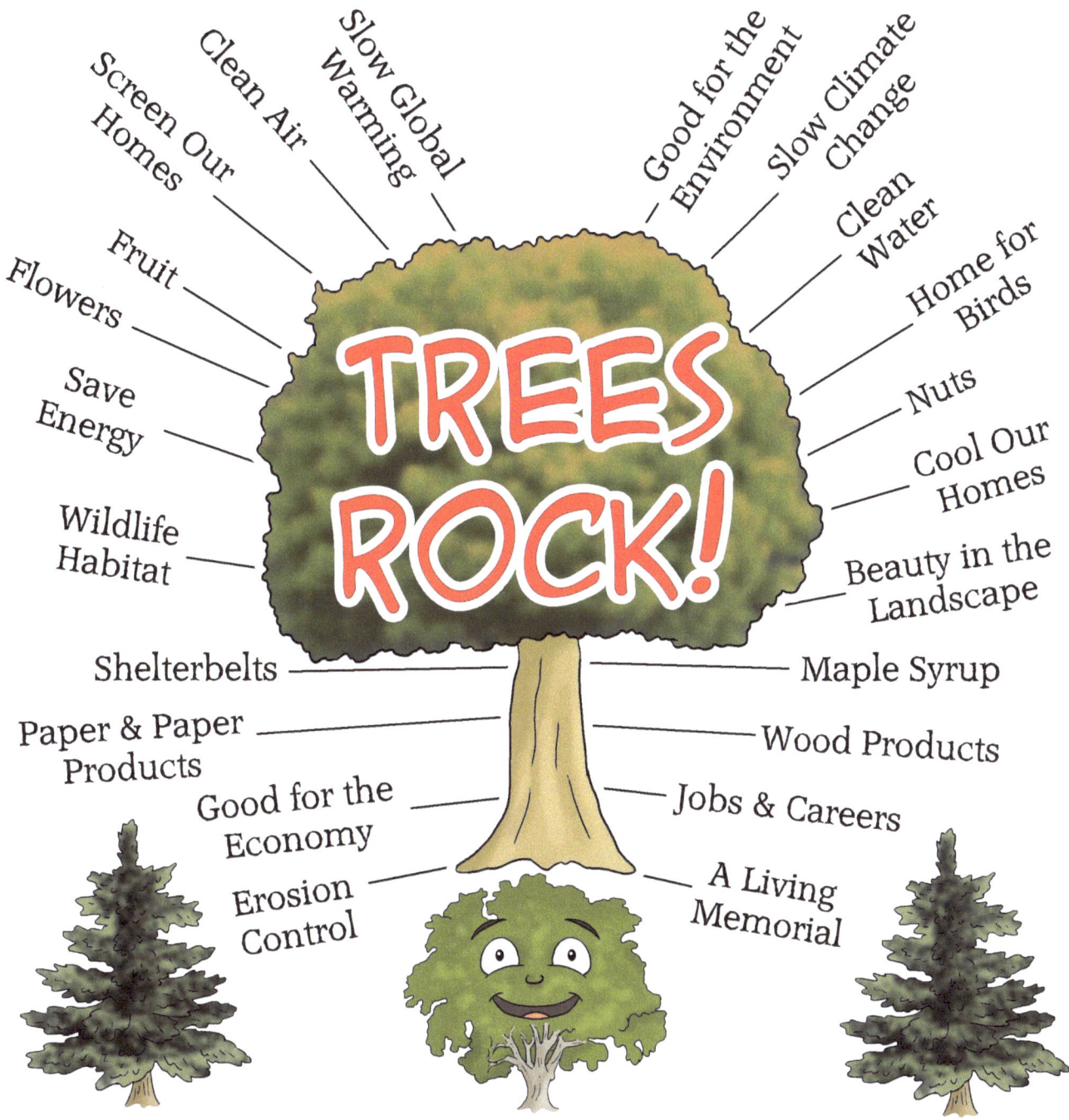

TREES ROCK!

Slow Global Warming
Clean Air
Screen Our Homes
Fruit
Flowers
Save Energy
Wildlife Habitat
Shelterbelts
Paper & Paper Products
Good for the Economy
Erosion Control

Good for the Environment
Slow Climate Change
Clean Water
Home for Birds
Nuts
Cool Our Homes
Beauty in the Landscape
Maple Syrup
Wood Products
Jobs & Careers
A Living Memorial

47

Worksheet: Discussion Questions

1. Name several kinds of trees you know about or like. Name some interesting things about them.

2. What are trees called that lose their leaves every year? What trees keep their needles all year long?

3. What is a plant Hardiness Zone? Why is it important to know a tree's or shrub's plant Hardiness Zone when planting a tree or shrub?

4. How does a seed develop, grow, and change to become a tree?

5. What are two different kinds of tree roots? What functions do these roots perform?

6. In a tree, what is the cambium layer? What is the function of the phloem? What is the function of the xylem?

7. What is the heartwood in the trunk? What function does it perform?

8. Describe photosynthesis? What happens during photosynthesis? Describe how a tree makes food in the leaves. What does a tree need to make food? What gas does it absorb through its leaves? What gas does a tree give off during photosynthesis?

9. What is the purpose or function of the bark of a tree?

10. Where is the crown of the tree? What is in the crown? What functions does the crown perform?

11. What are the four types of nursery stock that can be purchased at a nursery and garden center? Describe each type of stock. How are they different?

12. Describe pruning of different trees and shrubs, including timing and purposes for pruning.

13. Describe some ways of fertilizing trees, and three primary nutrients called macronutrients, including their functions, that help trees grow.

14. How do trees help the environment?

15. What jobs, occupations, and careers are involved with trees?

16. List some products made from wood.

Worksheet: Fill in the Blank

Word Bank

Xylem   Phloem   Roots   Trunk   Photosynthesis   Germination   Crown   Plant Hardiness Zone   Embryo
Heartwood   Stomata   Chlorophyll   Deciduous   Evergreen   Bark   Organic   Pruning   Seedlings
Cambium Cell Layer   Nitrogen   Phosphorous

1. Trees that keep their needles all year long: _____.

2. Trees that lose their needles each year: _____.

3. The _____is the climate zone in which a tree will grow best, based on climate and temperature.

4. The _____ is the very top of a tree.

5. The _____ is the main and longest part of a tree.

6. The _____ is inside the seed and contains cells that form tiny leaves, a stem, and a point that will become the root.

7. _____ occurs when the embryo splits the shell or outer coat of a seed.

8. _____ is a process in the leaves when a tree makes food. Leaves capture energy from the sun, create chlorophyll in the leaves, using water and taking in carbon dioxide, and giving off oxygen.

9. _____ is the green pigment found in chloroplasts in the leaves.

10. _____ are at the very bottom of a tree, absorb or take in water and nutrients for the tree, and help to hold up the tree and hold the soil in place.

11. _____ is the outermost layer of a trunk, stem, or branch and help to protect the tree injury, reduce water loss, and help to control the temperature inside the tree.

12. _____ is the layer where growing happens in a tree.

13. _____ is called the inner bark of a tree, on the inside of the bark, a vessel or tube like a pipeline. Food is passed downward and in other directions from the leaves through this tube.

14. _____ is a vessel or tube like a pipeline that carries water and minerals upward to the leaves. This tube is also called sapwood. This is new wood of a tree, or outer wood.

15. _____ is the centermost layer of a tree and is very strong. It contains no living cells.

16. _____ are small openings or pores on the undersides of leaves and other green plant parts, through which leaves absorb carbon dioxide, and through which oxygen and water vapor are

17. given off.

18. _____ is the mainly responsible for growth of leaves, and is very important in photosynthesis, green growth, flowering, and is often required in the greatest quantity.

19. _____ is mainly responsible for root and stem growth.

20. _____ are young plants grown from seeds.

21. _____ is material that comes from plant and animal sources.

22. _____ is a common procedure to improve a plant's growth, shape, and health, usually by removing specific branches or stems. This includes removal of weak, broken, dying, or dead branches.

About the Author

Roger F. Hartwich, Jr. M.S.E., M.S., B.S., B.A., has been a teacher and a landscaper/landscape designer/horticulturist/arborist for many years. Roger has been a teacher for many years in K-12 special education, elementary education, and all subject areas as a substitute teacher. He also has taught a course on the basics of home landscaping through a university extension service. Roger has owned/operated a horticultural services/landscaping and design company for many years, has worked as a horticulturist/head groundskeeper for a large institution, worked for tree nurseries and garden centers, and worked in financial services for a short period of time. Roger has strong interest in environmental preservation, trees and many other plants, horticulture, arboriculture, and landscape design as well as financial literacy.

Roger is an Army veteran and former Navy Reservist, United States Navy Retired, with 20 years of military service. Roger holds Masters' Degrees in Special Education and Recreation and Park Administration, a B.S. Degree in Elementary Education, a B.A. Degree in Social Science, German minor, and a technical college degree in horticulture/landscape technology and design.

Roger believes in relating and applying learning to real life skills. He has written this book for youth and adults about trees, based on fact, about tree anatomy, how trees live, grow, make food, and function, tree selection, planting, plant care and maintenance, and about their critical value and importance to people and society, the economy, environment, and birds and animals. It is hoped that this book will educate, inspire, and motivate youth and adults to become interested in trees, future tree planters, nurturers, and supporters of organizations involved with tree growing, reforestation, and planting in neighborhoods, town, cities, parks, and forests throughout the world.

Roger currently resides in Wisconsin.

www.ingramcontent.com/pod-product-compliance
Lightning Source LLC
Chambersburg PA
CBHW050911210326
41597CB00002B/92